CHEMICAL ENGINEERING

VITAMIN E AND
IONIZING RADIATION

CHEMICAL ENGINEERING METHODS AND TECHNOLOGY

Additional books in this series can be found on Nova's website under the Series tab.

Additional E-books in this series can be found on Nova's website under the E-book tab.

NUTRITION AND DIET RESEARCH PROGRESS

Additional books in this series can be found on Nova's website under the Series tab.

Additional E-books in this series can be found on Nova's website under the E-book tab.

CHEMICAL ENGINEERING METHODS AND TECHNOLOGY

VITAMIN E AND
IONIZING RADIATION

NELIDA L. DEL MASTRO

Nova Science Publishers, Inc.
New York

Copyright © 2010 by Nova Science Publishers, Inc.

All rights reserved. No part of this book may be reproduced, stored in a retrieval system or transmitted in any form or by any means: electronic, electrostatic, magnetic, tape, mechanical photocopying, recording or otherwise without the written permission of the Publisher.

For permission to use material from this book please contact us:
Telephone 631-231-7269; Fax 631-231-8175
Web Site: http://www.novapublishers.com

NOTICE TO THE READER

The Publisher has taken reasonable care in the preparation of this book, but makes no expressed or implied warranty of any kind and assumes no responsibility for any errors or omissions. No liability is assumed for incidental or consequential damages in connection with or arising out of information contained in this book. The Publisher shall not be liable for any special, consequential, or exemplary damages resulting, in whole or in part, from the readers' use of, or reliance upon, this material.

Independent verification should be sought for any data, advice or recommendations contained in this book. In addition, no responsibility is assumed by the publisher for any injury and/or damage to persons or property arising from any methods, products, instructions, ideas or otherwise contained in this publication.

This publication is designed to provide accurate and authoritative information with regard to the subject matter covered herein. It is sold with the clear understanding that the Publisher is not engaged in rendering legal or any other professional services. If legal or any other expert assistance is required, the services of a competent person should be sought. FROM A DECLARATION OF PARTICIPANTS JOINTLY ADOPTED BY A COMMITTEE OF THE AMERICAN BAR ASSOCIATION AND A COMMITTEE OF PUBLISHERS.

LIBRARY OF CONGRESS CATALOGING-IN-PUBLICATION DATA

Available upon Request
ISBN: 978-1-61761-191-9

Published by Nova Science Publishers, Inc. † *New York*

CONTENTS

PREFACE

Vitamin E - a family of eight natural structurally related tocopherols and tocotrienols compounds expressed generally as α-tocopherol - represents an essential component in human nutrition required for the preservation of lipids in stable form in biological systems and also in foods. In commonly consumed foods, vitamin E appears among the main antioxidants together with vitamins A and C and minerals like copper, zinc and selenium. The effect of gamma irradiation on the level of nutritional factors has been investigated. The role of reactive oxygen species in ionizing radiation injury and the potential of antioxidants to reduce these deleterious effects have also been studied by numerous groups in the last several decades. Naturally occurring antioxidants are considered able to behave as radioprotectors. Radiation protecting properties of vitamin E have been specifically described. The study of the effects of vitamin E on the formation of end products of radiation-induced free-radicals transformation has shown that the vitamin is able to oxidize α-hydroxyl-containing radicals, yielding the respective carbonyl or to reduce them to the initial molecules. On the other hand, the possibility of using gamma irradiation to improve the microbiological quality of different foods has been studied thoroughly. Food irradiation is presently applied commercially in the USA and France, among other countries. The need to eliminate bacterial pathogens from ready-to-eat food products must always be balanced with the maintenance of product quality. In addition to determining the effective ionizing radiation doses required for pathogen elimination and the effects of irradiation on product chemistry, nutritional value and sensory quality must also be determined. This review is an attempt to describe the state of the art of the role of vitamin E as a radioprotector as well as the behavior of vitamin E content in face of irradiation processing.

Chapter 1

INTRODUCTION

Tocopherol is any of several pale yellow fat-soluble oily liquid phenolic compounds that are derived from chroman and differ in number and location of methyl groups in the benzene ring (Webster, 1976). Vitamin E comprises a family of at least eight natural structurally related tocopherols (α-, β-, γ- and δ-) and tocotrienols (α-, β-, γ- and δ-) compounds generally expressed as α-tocopherol, since α-tocopherol is the compound demonstrating the highest vitamin E activity. This vitamin is only synthesized by plants. It is found in the dextrorotatory form especially in oil from seeds, in leaves and in fish-liver oils. It is available both in its natural form as RRR-α-tocopherol isolated from plant sources, but most commonly as synthetically manufactured all-rac-α-tocopherol. Synthetic all-rac-α-tocopherol consists of a racemic mixture of all eight possible stereoisomers. Assessing the correct biological activity through bioavailability and biopotency has been a great challenge during many years, as it is difficult to measure clinical endpoints in larger animals than rats and poultry.

The terms "vitamin E" and "α-tocopherol" are frequently used interchangeably in human nutrition, but it is imperative to distinguish between supplements of RRR-α-tocopherol or D-α-tocopherol and of synthetic all rac or DL-α-tocopherol because their biological activity is different. Alpha-tocopherol presents three quiral centers on the side chain making eight different stereoisomeric forms, whether R (right) or S (left) handed.

In terms of weight, natural vitamin E is considered about 36 percent more potent than the synthetic form of the vitamin. In fact, the "international unit," or IU, standard was developed to compensate for these differences. But there is no agreement about it. For instance, Yang et al. (2009) published some data showing that natural vitamin E was an effective vitamin E source but its relative bioavailability was substantially greater than 1:36 when compared with synthetic one.

Jensen and Lauridsen (2007) developed a relatively simple analytical method, which allowed to analyse the individual stereoisomers of α-tocopherol and to quantify the relative bioavailability of the individual stereoisomers. The method allowed the quantification of total tocopherol content and composition by normal phase HPLC and subsequent separation of the stereoisomers of α-tocopherol as methyl ethers by chiral HPLC. Using this method, the α-tocopherol stereoisomers were separated into five peaks.

The discussion on the bioavailability of RRR- and all-rac-α-tocopheryl acetate had primarily been based on human and animal studies using deuterium-labeled forms, whereby a higher biopotency of 2:1 (of RRR: all-rac) has been demonstrated, differing from the accepted biopotency ratio of 1.36:1. In agreement with previous studies, the 2S-forms played a very small role for the vitamin E activity due to their limited bioavailability (Jensen and Lauridsen, 2007).

Those mentioned authors found notable differences between animal species with regard to the biodiscrimination between the 2R-forms. Especially, cows preferentially transfer RRR-α-tocopherol into the milk and blood system. The distribution of the stereoisomer forms varies from tissue to tissue, and in some cases, higher levels of the synthetic 2R-forms than of the RRR-form are obtained, for example, for rats. However, the biodiscrimination of the stereoisomers forms was influenced by other factors such as age, dietary levels, and time after dosage. According to the referred authors, the focus should be on the bioactivity of the individual 2R-forms rather than just the comparison between RRR- and all-rac-α-tocopheryl acetate.

Although gastrointestinal absorption of all forms of vitamin E is supposed to be equivalent in man, the subsequent physiological steps are sharply in favor of the RRR form. This action is mediated by a cellular liver transfer protein that is specific for the RRR form of α-tocopherol. It maintains the plasma level by selectively choosing the RRR form and recycling it into plasma lipoproteins for distribution of the vitamin to every tissue and organ in the body.

In contrast to the tremendous recent advances in knowledge of vitamin E chemistry and biology, there is little evidence from clinical and epidemiologic studies on the beneficial effects of supplementation with vitamin E beyond the essential requirement (Meagher et al. 2001).

Most individuals are believed to obtain sufficient vitamin E from dietary sources, although individuals with very low-fat diets or intestinal malabsorption disorders may require supplementation. It was established, for instance, that for parenteral nutrition in cases of oxidative stress such as chronic inflammation, sepsis, lung distress syndrome and organ failure, there is a higher need for antioxidants intake (Biesalski, 2009).

Ames (2006) presented a substantial theory which postulates that natural selection favors short-term survival over long-term survival. He hypothesized that living organisms would achieve their short-term survival by prioritizing the allocation of scarce micronutrients. In that way, low micronutrient intake might accelerate the degenerative diseases of aging through allocation of scarce micronutrients by triage. According to this theory, age-related diseases like heart disease, cancer, and dementia may be unintended consequences of mechanisms developed during evolution to protect against episodic vitamin/mineral shortages. Indirectly, many studies strongly sustain this point of view and some of them will be mentioned in the next section.

The sources of vitamin E in the diet are primarily oils (soybean, corn, linseed, cotton, rapeseed, palm, sesame, wheat-germ, peanut, sunflower, olive), margarines (corn, soybean, sunflower), seeds (sesame, sunflower), nuts (almonds pecan, peanuts, Brazil nuts), cereal grains (corn, rice) and vegetables. The content of tocopherols in the vegetal kingdom depends on genetic variations but also on weather and climate in field-grown crops (Britz el al., 2008). Vitamin E can be also found in fatty tissues of selected animal sources (Scherz and Senser, 2000; USDA, 2006).

The regular daily requirement of vitamin E was established as about 15mg (Institute of Medicine, 2001, USDA, 2006). Dosing and daily allowance recommendations for vitamin E are often provided in α-tocopherol equivalents (ATE) to account for the different biological activities of the various forms of vitamin E, as well as in International Units (IU), which food and supplement labels often use. For conversion, 1mg ATE = 1.5 IU.

In a study to assess the content of α-, β-, γ- and δ-tocopherol and carotenoids in 10 different types of nuts, the authors found that the mean value of α-tocopherol equivalent ranged from non-detectable for macadamias to 33.1mg/100g in extracted oil from hazelnuts (Kornsteiner et al., 2006). The same authors found that among all nuts, almonds and hazelnuts had the highest mean α-tocopherols content (24.2 and 31.4mg/100g extracted oil, respectively), β- and γ-tocopherols were prevalent in Brazil nuts, cashew nuts, peanuts, pecans, pines, pistachios and walnuts. Chunhieng et al. (2008)

also found a high content of β-tocopherol in Brazil nuts. Other authors also found a predominance of α- and γ-tocopherols in seven nuts analyzed - with walnut and pecan oil containing the highest amounts - δ- and β-tocopherols also being detected in some samples, Brazil nut oil being the only significant source of δ-tocopherol (Miraliakbari & Shahidi, 2007). The results showed heterogenic amounts of this kind of antioxidants in nuts, emphasizing the importance of a mixed nuts intake.

Sundl et al. (2007), as well as, Hajimahmoodi et al. (2008) made a comparison of vitamin E content from composition tables and HPLC analysis and found substantial differences in calculated/measured vitamin E content of prepared foods: (i) between different food composition tables; (ii) between food composition tables and HPLC, and (iii) between different seasons. This can be explained by intrinsic variability (breeding, season, country of origin, ripeness, freshness) and food processing, as well as selection of food composition tables and should be taken into account when interpreting data of dietary intervention studies.

Franke et al. (2007) studied the levels of tocopherols and tocotrienols in 79 food items frequently consumed in Hawaii. They analyzed α-, β-, γ- and δ-tocopherol and α-, β-, γ- and δ-tocotrienol in addition to α -tocopheryl acetate. As compared to the few data available in the literature, their values agreed with some (corn flakes, mango fruit, fat-free mayonnaise, dry-roasted macadamia nuts, dry-roasted peanuts, mixed nuts, spaghetti/marinara pasta sauce, oils, and red bell pepper) but differed for many other items. These results provide information on the E vitamer content in foods, emphasize the vast differences of bioactivities of individual E vitamers, and confirm the need for analyses of foods consumed in specific given populations.

A study performed by Szymanska & Kruk (2008) was done to verify the literature data on atypical composition in photosynthetic tissues and isomers distribution in leaves of selected plant species. The average tocopherol level was in the range of 300-640μg/g fresh weight and accounted for 40-14-% of the chlorophyll amount. The isomers composition, usually dominated by α-tocopherol in leaves, was different in some plant species with high percentage of γ-tocopherol or δ-tocopherol or even exceptional absence of α-tocopherol, such is the case o *C. japonica*.

An in vitro study, followed up by a study with eight healthy men, found that a mixture of the carotenoids lycopene, β-carotene and lutein significantly reduced absorption of α-tocopherol, while vitamin C did not (Reboul et al., 2007). These authors also found that γ-tocopherol, also tended to reduce the absorption of the α-form. Then, multi-vitamin formulators may think carefully about formulations in order to maximize the nutritional quality of the supplements.

Chapter 2

ANTIOXIDANT AND OTHER
PROPERTIES OF VITAMIN E

Vitamin E is an important nutrient with antioxidant and non-antioxidant functions, so it is important to maintain an optimal vitamin E status for human health (Lodge, 2008). Antioxidant properties depend on the kind of oxidants as well as the reaction media. Ascorbate, α-tocopherol and β-carotene are traditionally considered very efficient antioxidants or electron donors (Machlin, 1991).

The fact that α-tocopherol possesses reducing properties enabled the development of a method to determine vitamin E based upon the reducing power of α-tocopherol against ferric chloride published by Emmerie A & Engel in 1938. The method used then was a colorimetric one based on the formation of ferrous salt, later determined with alpha-alpha' dipyridyl. Today traditional or new methods are also available (Anon., 1991; Ruperez et al., 1998; Amin, 2001; Sundl et al., 2007; Lodge, 2008; Cerretani et al., 2010).

The capacity of any nutrient to present in vitro antioxidant behavior is only an approximate reflection of their in vivo antioxidant effect, due to differences in antioxidant solubility/bioavailability within the digestive tract and the metabolism/conjugation of compounds (Fardet et al., 2008; McClement et al., 2008). Burton et al. (1983) had precisely described the power of vitamin E as an antioxidant in vitro and in vivo.

Packer et al. (1979) described that vitamin E (α-tocopherol) and vitamin C (ascorbic acid) react rapidly with organic free radicals, and recognized that their antioxidant properties were partially responsible for their biological activity (Slater, 1978). Tissue vitamin C levels are often considerably greater than those of vitamin E; for example, in liver the values are approximately 2 mM and 0.02 mM, respectively. Nevertheless, vitamin E is considerably more lipophilic than vitamin C, and in biomembranes it has been found to be the most potent antioxidant, particularly with respect to lipid peroxidation. Penetration to a precise site in the membrane may be an important feature of the protection against highly reactive radicals. It was suggested that the two vitamins act synergistically, vitamin E acting as the primary antioxidant and the resulting vitamin E radical then reacting with vitamin C to regenerate vitamin E. The work of Packer and colleagues reported the direct observation of this interaction by means of which the maintenance of vitamin E levels in tissues can be achieved.

It was shown that vitamin E acts as an antioxidant in oils since oxidation actively progressed as the vitamin E contents of the oils decreased in studies on grape seed oils (Kim et al, 2008). On the other hand, dietary α-tocopherol supplementation was found to reduce lipid oxidation indexes in fresh and sailed pork having no significant impact on color (Phillips et al., 2001).

The balance between the vitamin E and polyunsaturated fatty acid (PUFA) contents mainly determines the susceptibility to lipid peroxidation. Two examples from the literature can be mentioned on the subject.

Goffman & Bohme (2001) studied the storage stability of 30 hybrids of corn oil with various major fatty acids and tocopherols contents. They found that in corn oil where γ-tocopherol is predominant, this one was positively correlated with total tocopherols. Also, they found that a higher vitamin/PUFA ratio can be more easily achieved by increasing the vitamin E content than by modifying fatty acid profile.

Another different approach is described in the work of Villaverde et al. (2004). The authors aimed at clarifying whether dietary PUFA interferes with vitamin E absorption. They studied the effect of dietary vitamin E inclusion level (0 - 400 mg/kg) and the degree of fatty acid unsaturation (15 - 61 g PUFA/kg) on vitamin E apparent absorption and tissue deposition in poultry. Increasing dietary levels of vitamin E reduced its apparent absorption. The more saturated diet reduced fat and vitamin E apparent absorption while PUFA levels from 34 to 61 g/kg did not modify this parameter but reduced the hepatic vitamin E concentration, suggesting a greater systemic use of the vitamin. These results suggest that PUFA do not limit vitamin E absorption, although they may increase its degradation in the gastrointestinal tract.

From studies of sex and age dependence of developmental changes in hepatic antioxidant capacity as a whole, it was established that males had lower superoxide dismutase and vitamin E and might have lower glutathione reductase, while females showed less cytosolic glutathione-S-transferase.

Hepatic antioxidants are high in neonates, decline throughout childhood, and then increase in adolescence to adult levels according to Miyagi et al., (2009). In a study with laboratory animals, vitamin E was measured in rats that had previously received one oral dose of this vitamin. Whereas vitamin E content in adipose tissue did not differ between late-pregnant and virgin rats, it was significantly higher in mammary glands of pregnant rats, and this difference could be related to the enhanced lipoprotein lipase activity in this group (Ruperez et al., 1998).

The findings of Kim et al. (2010) support a role of antioxidant vitamin (A, C, E and β-carotene) intake in the decrease of cervical cancer risk, one of the most common gynecological malignancies.

Epidemiological or cell lines studies have shown an inverse relationship between nut intake or tocopherol presence and chronic diseases such as cardiovascular diseases and some kinds of cancer (Lim et al., 2009; Roche et al., 2009). The consumption of Brazil nuts, for instance, rich in dietary antioxidants, possessing phenolics and flavonoids in both free and bound forms and particularly rich in tocopherols, phytosterols and squalene was found to present several such associated health benefits (Yang, 2009).

Human intake of fruits and vegetables, vitamin C, E and fiber was negatively associated with high-sensitivity C-reactive protein (Oliveira et al., 2009). Also, some recent studies suggest that increased intakes of beta-carotene, vitamin E, folic acid, and iron may reduce the risk of atopic dermatitis, a form of eczema atopic dermatitis in young children (Oh et al. (2010).

Vitamins C and E are able to reduce oxidative stress, improve vascular function and structure, and prevent progression of hypertension in adult stroke-prone spontaneously hypertensive rats according to the work of Chen et al (2001). The authors considered that these effects may be mediated via modulation of enzyme systems that generate free radicals.

Capuron et al. (2009) showed that old patients with poor physical health status, as assessed by the Medical Outcomes Study 36 Items Short-Form Health survey (SF-36), exhibited lower circulating concentrations of α-tocopherol together with increased concentrations of inflammatory markers. Similarly, poor mental health scores on the SF-36 were associated with lower concentrations of α-tocopherol and decreased concentrations of tryptophan.

Vitamin E was found to have an influence on glycemic parameters in diabetic rats and neuroprotective effect on the total myenteric population, while the morphometric characteristics of the intestinal wall were unaffected (Roldi et al, 2009).

Vitamin E succinate was reported to reduce oxidative stress and improve mitochondrial function in mice with steatotic livers before and after ischemia/reperfusion (Evans et al., 2009).

Shireen et al. (2008) presented some results that clearly indicated that vitamin E and C, both separately and together, increased antioxidant enzyme activities in liver and muscles of rats. However, a combination of vitamin E and C enhanced antioxidant enzyme activity more significantly, suggesting the possible role of vitamin C and E and their combination in reducing the risk of chronic diseases related to oxidative stress.

In a study performed by Haflah et al. (2009) it was found that oral palm vitamin E in a dose of 400mg daily taken had the potential to reduce symptoms of patients with osteoarthritis of the knee, being as effective as glucosamine sulphate and free of serious side effects.

A high dose of vitamin E supplementation of 2,000IU per day has been shown to delay Alzheimer's disease progression without increasing mortality (Pavlik et al., 2009).

As oxidative stress may influence vitamin E status, some studies comparing vitamin E biokinetics and metabolism in cigarette smokers and non-smokers have been able to show differences in vitamin E processing in smokers (Lodge, 2008).

Vitamin E was found to partially prevent from 30 to 50% of toxic effects induced by zearalenone, a non-steroidal estrogenic mycotoxin (Ouanes et al., 2003). In this experiment genotoxicity tests were applied either in cultured Vero cells or in bone marrow cells extracted from mice to which vitamin E had been previously administered.

The abundance of α-tocopherol in the human body led biologists to discard the non-tocopherol vitamin E molecules as topics for basic and clinical research (Hensley et al., 2004). Some developments warrant a serious reconsideration of this conventional wisdom. Early studies conducted between 1922 and 1950 indicated that α-tocopherol was specific among the tocopherols in allowing fertility of laboratory animals. New and unexpected biological activities have been reported for the desmethyl tocopherols, such as γ-tocopherol, and for specific tocopherol metabolites, most notably the carboxyethyl-hydroxychroman (CEHC) products. The activities of these other tocopherols do not map directly to their chemical antioxidant behavior but rather reflect anti-inflammatory, antineoplastic, and natriuretic functions possibly mediated through specific binding interactions. Moreover, a nascent body of epidemiological data suggests that γ-tocopherol is a better negative risk factor for certain types of cancer and myocardial infarction than α - tocopherol is. The potential public health implications are enormous, given the extreme popularity of α–tocopherol supplementation which can unintentionally deplete the body of γ -tocopherol. These findings may or may not signal a major paradigm shift in free radical biology and medicine. The data argue for thorough experimental and epidemiological reappraisal of desmethyl tocopherols, especially within the contexts of cardiovascular disease and cancer biology (Chandran et al., 2006).

According to Wagner et al. (2004) the data available in scientific literature provides evidence that γ-tocopherol may be an important player in human health and might synergize α-tocopherol in protection against chronic diseases. Even though γ-tocopherol is not retained and distributed in the body in the same way as α-tocopherol, its biological effects, especially detoxification of reactive nitrogen species, its anti-inflammatory activity, and its effect in reducing risks of cardiovascular disease and cancer are already shown. Nevertheless, more studies must be conducted to clarify why the body apparently utilizes α-tocopherol preferentially, and what functions other forms of vitamin E have.

Yoshida et al. (2007) established that the α-isomers of both tocopherol and tocotrienol are the most effective scavengers in vitamin E. However, the tocotrienols appeared to be more readily incorporated into liposomal membranes, which suggests that this group of molecules would be most effective in protecting against cell damage by reactive species.

Tocotrienols possess powerful neuroprotective, anti-cancer and cholesterol lowering properties that are often not exhibited by tocopherols. Chandran et al., (2006) described that some developments in vitamin E research clearly indicated that members of the vitamin E family were not redundant with respect to their biological functions and individual isomers have different propensities with respect to these novel, nontraditional roles. Last year, some authors presented evidences recognizing that nanomolar vitamin E α-tocotrienol inhibits glutamate-induced activation of phospholipase A2 and causes in that way a neuroprotection (Khanna et al., 2009).

As Schneider (2005) had corroborated, vitamin E, although initially seen as nature's most potent lipid-soluble antioxidant playing a crucial role in mammalian reproduction, later was recognized to have many more facets, depending on the physiological context. Although mainly acting as an antioxidant, vitamin E can also be a pro-oxidant; it can even have nonantioxidant functions: as a signaling molecule, as a regulator of gene expression, and, possibly, in the prevention of cancer and atherosclerosis. Alpha-tocotrienol, γ-tocopherol, and δ-tocotrienol have emerged as vitamin E molecules with functions in health and disease that are clearly distinct from that of α-tocopherol. At nanomolar concentration, α-tocotrienol, not α-tocopherol, prevents neurodegeneration. On a concentration basis, this finding represents the most potent of all biological functions exhibited by any natural vitamin E molecule.

As an expanding body of evidence supports that members of the vitamin E family are functionally unique, title claims in manuscripts should be limited to the specific form of vitamin E studied. For example, evidence for toxicity of a specific form of tocopherol in excess may not be used to conclude that high-dosage "vitamin E" supplementation may increase all-cause mortality. Such conclusion incorrectly implies that tocotrienols are toxic as well under conditions where tocotrienols were not even considered. The current state of knowledge warrants strategic investment into the lesser known forms of vitamin E. This will enable prudent selection of the appropriate vitamin E molecule for studies addressing a specific need.

The particular beneficial effects of the individual isomers have to be considered when dissecting the physiological impact of dietary vitamin E or supplements (mainly containing only the α-tocopherol isomer) in clinical trials. These considerations are also relevant, for instance, for the design of transgenic crop plants with the goal of enhancing vitamin E content because an engineered biosynthetic pathway may be biased toward the formation of one isomer.

A study on maternal α-tocopheryl acetate and post weaning vitamin C supplementation suggested that these substances can be used as a nutritional strategy for increasing α-tocopherol status and immune responses of weaned piglets (Lauridsen & Jensen, 2005).

Both γ-tocopherol and a potent antitumor analog of vitamin E, the α-tocopherol ether-linked acetic acid analog exhibited promising anticancer properties in vivo and in vitro, whereas the manufactured synthetic form of vitamin E, all-racemic-α-tocopherol exhibited promising anticancer properties in vivo only. Also, not only has α-tocopherol failed to exhibit anticancer properties but it also reduced anticancer actions of γ-tocopherol in vivo and γ-tocopherol and α-tocopherol analog in vitro (Yu et al, 2009).

Miyazaki et al. (2008) reported that impaired graft healing due to hypercholesterolemia is prevented by dietary supplementation with α-tocopherol. Toumpanakis et al, (2009) described that in healthy subjects, antioxidants like vitamins E, C and A differentially affected monocytes with a general tendency of augmenting of their cytokine production.

There are also reports from the literature where vitamin supplements failed to show health benefits. To explore the association of supplemental multivitamins, vitamin C, vitamin E, and folate with incident lung cancer,

Slatore et al, (2008) made a prospective cohort of 77,721 men and women aged 50–76 study.Cases were identified through the Seattle–Puget Sound SEER (Surveillance, Epidemiology, and End Results) cancer registry. In the conditions of their study, long-term use of supplemental multivitamins, vitamin C, vitamin E, and folate did not show to reduce the risk of lung cancer.

Atay el al. (2009) suggested that there are some improving effects of dietary vitamin E supplementation on fattening performance of lambs.

A study on maternal all rac-α-tocopheryl acetate and post weaning vitamin C supplementation suggests a nutritional strategy for increasing α-tocopherol status and immune responses of weaned piglets (Lauridsen & Jensen, 2005).

In vivo analysis revealed that both serum lipid peroxides and splenocyte peroxides were found to be reduced with increased dietary vitamin E levels in a dose-dependent manner, whereas, serum vitamin E levels were found to be augmented with increased levels of dietary vitamin E, according a work performed by Avula & Fernandes (2000) in male mice. The results of their work indicated that dietary vitamin E seemed to alter both cell proliferation and programmed cell death (apoptosis vs. necrosis) by lowering lipid peroxides.

Uysal et al. (2009) reported that the application of α-tocopherol to orthopedically expanded inter-premaxillary suture areas during early stages might stimulate bone formation and shorten the retention period in rats.

Giardini et al. (1984) have established the effects of α-tocopherol administration on red blood cell membrane lipid peroxidation protection in hemodialysis patients. Lucio et al. (2009) also described some aspects related to the protection of membranes lipoperoxidation by vitamin E and trolox.

Plasma vitamin E status is disturbed as part of systemic inflammatory response and although the value of correcting vitamin E concentrations by lipids is well established in population studies, the actual vitamin E status should be assessed using red blood cell α-tocopherol measurements according an study published by Vasilaki et al. (2009).

Recent work suggests that vitamin E plays an important role in immunoregulatory processes in patients with Sjogren's syndrome (Szodoray et al., 2010).

Vitamin E supplementation might transiently increase tuberculosis risk particularly in males who smoke heavily and have high dietary vitamin C intake according to a survey made by Hemila & Kaprio (2008), although vitamin E supplementation alone had no overall effect on the incidence of tuberculosis.

Chapter 3

IONIZING RADIATION

Ionizing radiation is radiation from both natural and man-made sources which is energetic enough to produce ions from atoms or molecules by removing electrons. Radiation processes include the use of accelerated electrons or photons, ranging from radio-frequencies, microwaves, infra-red radiation, light, ultraviolet radiation, X- and γ rays. When accelerated electrons and short wavelength (4.1×10^{-3} nm) X- and γ-ray (1.0×10^{-3} nm) are used photons interact with matter at the atomic level. The raising of an electron in an atom or molecule to a higher energy level without actual ejection is called excitation. If the radiation has sufficient energy to eject one or more orbital electrons from the atom or molecule, the process is called ionization and that radiation is said to be ionizing radiation (Hall, 1994).

The interaction of high-energy radiation or ionizing radiation with various forms of matter has been shown to produce a variety of chemical effects, depending upon the system. These effects fall into various classes, among which are oxidation and reduction, polymerization and depolymerization and or cross-linking of polymeric materials (Lawton et al., 1953).

The amount of radiation energy absorbed is measured in units of grays where one gray (Gy) equals one joule per kilogram. When molecules absorb ionizing energy they form ions or free radicals. A free radical contains an unpaired electron in the outer shell, as a result of which it is highly reactive, that react to form stable radiolytic products (Woods & Pikaev, 1994).

Ionizing irradiation treatment has many applications for human benefit, whether in medicine as therapy and diagnosis, or in industry to enhance physical and chemical properties of materials. In medical practice, ionizing radiation sources are mainly used to take advantage of the radiation capacity of killing cells like tumor cells. Among studies conducted to improve the understanding of how cells respond to radiation and how defects in those responses can lead to mutations that cause cancer, the work from UK and Denmark (Golberg et al., 2003; Chapman & Jackson, 2008)can be mentioned.

Currently, microorganisms cause millions of cases of foodborne illnesses and thousands of deaths on a global basis. Irradiation treatment of foods can be used to eliminate or reduce undesirable contaminants, such as pathogens, spoilage microorganisms, insects, moulds and bacteria or toxic by-products. For that reason, a lot of experimental studies were performed all around the world on food irradiation (Farkas, 1989; Heide & Boegl, 1990; Thayer, 1990: Clavero et al., 1994; WHO, 1994; Diehl, 1995; Crawford & Ruff, 1996; Mastro, 1999a; Mastro, 1999b; Bosch, 2005; Kume et al., 2009).

Irradiation of medical devices and disposables has a long history of use (Derr, 1993) but although the irradiation process is a particularly effective method for the preservation of food for sanitary or phytosanitary reasons, it is applied in few countries which have proper legislation (Kume et al., 2009). Then, the acceptance of irradiated foods is not worldwide and few countries produce commercially irradiated food items, among them, the USA and France. Spices are the most commonly irradiated food. Other commercially-available irradiated foods include a variety of fruits and vegetables, rice, potatoes, onions, sausage and muscle food like meat, poultry and seafoods (Olson, 1998).

Food irradiation can also be used in some market niches to provide sterilized foods for astronauts, some hospitalized patients or radical sport practitioners (Mastro & Mattiolo, 2010). The irradiation of some food items can be used also to overcome disadvantages of some food products. For instance, a work described by Shin el al. (2008) showed the reduction of the antigenicity of milk protein contained in powdered milk by γ-irradiation.

Chapter 4

RADIOPROTECTIVE ROLES
OF VITAMIN E

Pharmacologic approaches of protection against radiation-induced damage were described by Weiss (1997). Protectors generally are classified as either sulfhydryl compounds, other antioxidants, or receptor-mediated agents (e.g., bioactive lipids, cytokines, and growth factors). The potential of antioxidants to reduce ionizing radiation injury produced by reactive oxygen species has been studied in animal models for several decades (Weiss & Landauer, 2000). The mechanism of action of radioprotective compounds involves generally the scavenging of free radicals (Hall, 1994). There are some similarities among the defense against ionizing radiation and other situations which depend in some stage on the production of free radicals. For that reason some antioxidants or radioprotectors can act also to prevent damages coming from toxic chemicals (Ouanes et al., 2003; Agarwal et al., 2010).

As it was correctly pointed out (Riley, 1994) the most important electron acceptor in the biosphere is molecular oxygen which, by virtue of its bi-radical nature, readily accepts unpaired electrons to give rise to a series of partially reduced species collectively known as reactive oxygen species (ROS). These include superoxide ($O\bullet^{-2}$), hydrogen peroxide (H_2O_2), hydroxyl radical ($HO\bullet$) and peroxyl ($ROO\bullet$) and alkoxyl ($RO\bullet$) radicals which may be involved in the initiation and propagation of free radical chain.

ROS are formed enzymatically, chemically, photochemical photochemically, and by irradiation of biological material. Organisms living in an aerobic environment were forced to evolve effective cellular strategies to detoxify ROS. The adaptive responses, especially those to radiation, are defensive regulation mechanisms by which oxidative stress (conditioning irradiation) elicits a response against damage because of subsequent stress (challenging irradiation) as described by Miura (2004).

Natural mechanisms of living organisms to restrict and control such processes include antioxidant defenses such as chain-breaking antioxidant compounds capable of forming stable free radicals (e.g. ascorbate, α-tocopherol) and the evolution of enzyme systems (e.g. superoxide dismutase, catalase, peroxidases) that diminish the intracellular concentration of the ROS. Although some ROS perform useful functions, the production of ROS exceeding the ability of the organism to mount an antioxidant defense results in oxidative stress and the ensuing tissue damage may be involved in certain disease processes.

Evidence that ROS are involved in primary pathological mechanisms is a feature mainly of extraneous physical or chemical perturbations of which radiation is perhaps the major contributor (Gate et al., 1999). One of the important radiation-induced free-radical species is the hydroxyl radical which indiscriminately attacks neighboring molecules often at near diffusion-controlled rates. Hydroxyl radicals are generated by ionizing radiation either directly by oxidation of water, or indirectly by the formation of secondary partially ROS. These may be subsequently converted to hydroxyl radicals by further reduction ('activation') by metabolic processes in the cell. Secondary radiation injury is therefore influenced by the cellular antioxidant status and the amount and availability of activating mechanisms. The biological response to radiation may be modulated by alterations in factors affecting these secondary mechanisms of cellular injury.

Antioxidants are supposed to quench oxidation by transferring hydrogen atoms to free radicals. Naturally occurring antioxidants, such as vitamin E and selenium, were found to be less effective radioprotectors than synthetic thiols but may provide a longer window of protection against lethality and other effects of low dose, low-dose rate exposures. Many natural antioxidants have antimutagenic properties that need further examination with respect to long term radiation effects. Modulation of endogenous antioxidants, such as superoxide dismutase, may be useful in specific radiotherapy protocols (Weiss & Landauer, 2000).

There are many examples in the literature about the involvement of vitamin E in the process of radiation defense. Some thirty years ago the work of Oski (1980) showed the capacity of vitamin E against free radicals.

Borek et al. (1986) published some results showing that selenium and vitamin E were able to inhibit radiogenic and chemically induced transformation in vitro via different mechanisms.

In 1990 some experiments performed in our laboratory highlighted the role of α-tocopherol, as well as peanut oil in the defense against ionizing radiation damage in mice (Mastro & Villavicencio, 1990). In the same year Minkova & Pantev (1990) published the results of their work showing the protection against radiation effects by the combined pretreatment administration of α-tocopherol, anthocyans and pyracetam. Minkova et al. (1990) also described the radioprotective action of α-tocopherol when administered together with other substances before irradiation in adult male mice. Another group also found that vitamins C, E and β-carotene reduced DNA damage before, as well as after γ-ray irradiation of human lymphocytes in vitro (Konopacka & Rzeszowska-Wolny, 2001).

Srinivasan and Weiss (1992) determined that injectable vitamin E administered alone or with WR-3689 enhanced the survival of irradiated mice. In the same year, Empey et al. (1992) reported the mucosal protective effects of vitamin E and misoprostol during acute radiation-induced enteritis in rats.

Sarma et al. (1993) reported the protective effects of vitamin C and E against γ-rays-induced chromosomal damage in mouse.mice.

Intestinal radioprotection by vitamin E (α-tocopherol) was demonstrated by Felemovicius et al. (1995). They used the small intestine of a rat as a study system. Crypt cell numbers, mucosal height, and goblet cell numbers were significantly protected from radiation effects by dietary α-tocopherol pretreatment and by lumenal application of the vitamin. The study indicated that vitamin E can serve as a partial protection against acute irradiation enteritis, whether given as a chronic oral systemic pretreatment or as a brief topical application.

Yoshimura et al. (2002) examined the effects of dietary vitamin E on oxidative damage to DNA and lipids in the liver a few days after total body irradiation in ODS rats. The animals, which lack vitamin C synthesis, were fed either a low vitamin E diet (4.3 mg/kg) or a basal diet (75.6 mg/kg) for 5 weeks while vitamin C was supplied in the drinking water. The vitamin E level in the liver of the low vitamin E group was lower and the levels of lipid peroxides were higher compared to those of the basal vitamin E group. The level of 8-hydroxydeoxyguanosine (SOHdG), a marker of oxidative DNA damage, did not differ between the low and the basal vitamin E groups. When the rats received total body irradiation at the dose of 3 Gy and were killed on day 6, the levels of markers of oxidative DNA damage increased in the low vitamin E group, but not in the basal vitamin E group. The concentrations of vitamin C and glutathione in the liver did not differ between the two groups. These results suggest that dietary vitamin E can prevent the oxidative damage to DNA and lipids in the liver which appear a few days after total body irradiation at dose of 3 Gy.

It is known that superoxide slowly and irreversibly oxidizes tocopherol in organic solvent and produces tocopheryl radical, but the reaction is insignificant in aqueous solution as reported by Halliwell & Gutteridge (2001). They showed also that tocopherols react irreversibly with singlet oxygen and produce tocopherol hydroperoxydienone, tocopherylquinone, and quinone epoxide the reaction rate of tocopherol with singlet oxygen being affected by the structure of tocopherol.

The protective effects of DL-α-tocopherol acetate and sodium selenate on the liver of rats exposed to gamma radiation were studied by Yanardag et al. (2001). Based on morphological and biochemical observations it was concluded that the ip administration of DL-α-tocopherol acetate and sodium selenate exerts a protective effect against liver radiation damage.

The radioprotection conferred by a highly namely, 2-(α-D-glucopyranosyl) methyl-2,5,7,8-tetramethylchroman-6-ol (TMG) in *Saccharomyces cerevisiae* was studied by Singh et al. (2001). Cells grown in standard YEPD-agar medium and irradiated in the presence of TMG showed a concentration dependent higher survival up to 10 mM of TMG in comparison to cells irradiated in distilled water. Treatment of TMG to cells given either before or immediately after irradiation but not during irradiation, had no effect on their radiation response. *S. cerevisiae* strain LP 1383 (rad52) which is defective in recombination repair showed enhanced radioresistance only when subjected to irradiation in presence of TMG. These authors concluded that TMG confers radioresistance in *S. cerevisiae* possibly by two mechanisms viz. (i), by eliminating radiation induced reactive free radicals when the irradiation is carried out in the presence of TMG and (ii), by activating an error prone repair process involving RAD5 RAD52 gene, when the ells are grown in the medium containing TMG.

Another group also used, in their study, the water soluble glucose derivative of α-tocopherol, TMG. They concluded that the high water solubility and effectiveness when administered post-irradiation in adult Swiss albino mice favor TMG as a likely candidate for protection in case of accidental exposures (Satyamitra et al. (2001).

Vitamin E showed to be an effective antioxidant during γ-irradiation even in non biological media. The oxidation of high density polyethylene (HDPE), both unstabilised and vitamin E stabilised, has been studied by infrared (IR) and electron paramagnetic resonance (EPR) spectroscopies in the period following γ-irradiation at doses from 1 to 60 kGy (range of food sterilisation). Vitamin E, although an effective antioxidant during γ-ii..   ..ion, was less effective in reducing the post-irradiation changes (Mallegol et al., 2001).

Studies in vitro using *E. coli* bacteria (AB 1157) as a model for living systems were performed with vitamin E acetate, β-carotene and vitamin C in order to investigate their radiation protecting ability (Kammerer et al., 2001). A cobalt-60 commercial irradiator providing a dose rate of 60 Gy/min was used as γ-rays irradiation source. The referred substances in concentration of 10 mol dm^{-3} showed to possess very efficient radiation protecting properties, in addition to their well known antioxidant action. The strongest radiation protecting effect was observed for vitamin C ($\Delta D_{37}=90$), followed by vitamin E acetate ($\Delta D_{37}=60$) and β-carotene ($\Delta D_{37}=50$). The authors reckoned that the radiation protecting mechanism of the three vitamins is not completely known, but it is very likely to be due to both: electron donor ability to convert oxidizing species (e.g. OH•, HO2• /O2•, ROO•, etc.) in none or less reactive anions as well as by their free radical scavenging properties.

Manzi el al. (2003) evaluated the action of the vitamin E as a radioprotective agent in the process of tissue reparation in rats. The mammals were submitted to a surgical procedure, which consisted of a wound done in the fore dorsal area. The animals were divided into five groups: group C (controls) - wound; group VE - previous treatment with vitamin E (90 UI); group IR - wound and irradiation of the borders three days after surgery; group VEIR - previous treatment with 90 UI of the vitamin E and irradiation of the borders three days after the surgery; group OIR - previous treatment with olive oil and irradiation of the borders three days after surgery. The radioprotective effect of the vitamin E was evaluated using hematoxylin-eosin stained specimens in order to identify granulation tissue, at 4, 7, 14 and 21 days after the surgical procedures. The results showed that 6 Gy of electron irradiation with a beam of 6 MeV caused retardation of the tissue repairing process and that vitamin E was effective as a radioprotective agent.

Shadyro et al. (2005) studied the effects of vitamin B, C, E, K as well as coenzymes Q on the formation of end products of radiation-induced free-radical transformation of ethanol, ethylene glycol, α-methylglycoside and glucose in aqueous solutions. They concluded that vitamins and coenzymes interact with α-hydroxyl-containing radicals. In the presence of those substances, recombination reactions of α-hydroxylalkyl radicals and fragmentation of α-hydroxyl-β-substituted organic radicals were suppressed. They ascribed the observed effect to the ability of the vitamins and coenzymes to either oxidize α-hydroxyl-containing radicals yielding the respective carbonyl compounds or to reduce them into the initial molecules.

As there is some concern about the fact that the free radical scavenging effect of antioxidants added to food might reduce the antimicrobial effectiveness of ionizing radiation, Romero and Mendonça (2005) performed the following study. Raw ground turkey breast meat from birds fed diets containing 0 (control), 50, 100 and 200 IU kg^{-1} of vitamin E was inoculated with a five-strain mixture of L. monocytogenes to give approximately 10^7 CFU g^{-1}. Inoculated samples were irradiated up to 2kGy and stored aerobically (12 days) or under vacuum(42 days) at 4°C. L. monocytogenes survivors were determined by plating and colonies were counted. Irradiation at 2kGy resulted in an approximate 3.5-log reduction of initial numbers of L. monocytogenes. There were no significant differences in D-values (decimal reduction times) for the microorganisms in meat irrespective of vitamin E treatment. Then, vitamin E treatment did not compromise the microbial safety of the irradiated product and presented improvement in color stability.

Choe and Min (2005) established the possible role of tocopherols in the interaction with ionizing radiation. They described that reactions of ROS with food components are able to produce undesirable volatile compounds, can destroy essential nutrients, and can change the functionalities of proteins, lipids, and carbohydrates. According to them, lipid oxidation by ROS produces low-molecular-weight volatile aldehydes, alcohols, and hydrocarbons. Also, ROS would cause crosslink or cleavage of proteins and would produce low-molecular-weight carbonyl compounds from carbohydrates. They supposed that vitamins would be easily oxidized by ROS, especially singlet oxygen as the singlet oxygen reaction rate is the highest in β-carotene followed by tocopherol, riboflavin, vitamin D, and ascorbic acid.

The above referred authors formulated the following hypothesis: tocopherols react with peroxyl radicals. Hydrogen atom at 6-hydroxy group on chromanol ring of tocopherols is transferred to peroxyl radical producing alkyl hydroperoxide and tocopheryl radical, which is relatively stable due to resonance structure. Reaction of tocopherols with peroxyl radicals slows down the lipid oxidation. Tocopherols would compete with unsaturated lipids for lipid peroxyl radicals. Lipid peroxyl radical with the reduction potential of 1000 mV reacts with tocopherol much faster at 104 ~109/M/s than with lipid at 10~ 60/M/s due to lower reduction potential of tocopherols of 500 mV than that of unsaturated lipid of 600 mV. One tocopherol molecule could protect about 103~108 polyunsaturated fatty acid molecules at low peroxide. Even though it is very slow, the tocopheryl radical sometimes can abstract hydrogen from lipids to give tocopherol and lipid radical at very low concentration of lipid peroxy radical. Lipid radical could promote the lipid oxidation, the so-called tocopherol-mediated peroxidation (Bowry and Stocker 1993. Ascorbic acid can quickly reduce tocopheryl radical and prevent tocopherol mediated peroxidation. Then tocopherol can be slowly oxidized by hydroperoxyl radicals at the rate of $2.0 \times 105/M/s$ and produce substituted tocopherones and epoxy or hydroperoxy tocopherone (Halliwell and Gutteridge 2001).

As mentioned by Fryer (2006), vitamin E (α-tocopherol) is essential for the prevention of photo-oxidative deterioration of biomembranes, serving as a chain-breaking antioxidant, protecting cell membranes against free-radical damage. The author put in relief this fact in connection with the occurrence, relative biological activity and distribution of tocopherols in photosynthetic membranes together with the possible biochemical and biophysical mechanisms by which tocopherols confer protection upon illuminated membranes. He also stressed the common protective effects of vitamin E in photo-synthetic membranes and in medically important light-induced diseases and conditions of the skin and eye in animal cell membranes. In particular, he emphasized the importance of the vitamin E-vitamin C thylakoid antioxidant system in terms of susceptibility to photo-oxidative damage under stress conditions including chilling, ageing and senescence, drought, atmospheric pollutants, herbicides and photosensitizing fungal toxins.

Sezen et al. (2008) determined the effects of vitamin E and l-carnitine supplementation, individually or in combination, on radiation-induced brain and retinal damages in a rat model. Group 1 received no treatment (control arm). Group 2 received a total dose of 15Gy external radiotherapy (RT) to whole brain by cobalt-60 teletherapy machine. Groups 3, 4, and 5 received irradiation plus 40 kg^{-1} day^{-1} Vitamin E or 200 mg $kg^{-1}day^{-1}$ l-carnitine alone or in combination. Brain and retinal damages were histopathologically evaluated by two independent pathologists. Antioxidant enzyme levels were also measured. Radiation significantly increased brain and retinal damages. A significant increase in malondialdehyde levels as well as a decrease in superoxide dismutase and catalase enzymes in brain was found in group 2. Separate administration of Vitamin E+RT and l-carnitine+RT significantly reduced the severity of brain and retinal damages and decreased the malondialdehyde levels and increased the activity of superoxide dismutase and catalase enzymes in the brain. However, the combined use of Vitamin E and l-carnitine plus irradiation interestingly did not exhibit an additive radioprotective effect.

Increased intakes of vitamins C and E and other antioxidants from the diet were shown to protect against DNA damage in people exposed to ionizing radiation such as pilots (Yong et al., 2009).

Chapter 5

STABILITY OF VITAMIN E CONTENT

As it was presented before, vitamin E acts as a cell protector in acute oxidative stress conditions and diseases and also against radiation-induced damage in different systems. Vitamin E being a free radical scavenger there is a growing concern that irradiation might reduce their content on food products prepared with ingredients rich in any of the dietary source of the vitamin.

Traditionally, of all the fat soluble vitamins, (Knapp & Tappel, 1961) vitamin E was considered to be the most sensitive to gamma radiation. Diehl (1979) suggested that irradiation would promote autoxidation from the unsaturated fatty acids along with the production of peroxide and other reaction products. Nevertheless, the effects of irradiation on the vitamin E level in foods, particularly after processing, are complex because multiple reactions due to irradiation and processing occur simultaneously (Thorne, 1991).

A comprehensive survey about food vitamin E content under the effects of heating treatments was conducted by Leskova et al. (2006). They concluded that it is difficult to determine the most aggressive culinary method to affect vitamin E. Likely, the most common heat treatments, such as broiling or roasting, caused a high loss of the nutrient. The vitamin E content in food treated in vegetable oil increased or remained stable because vegetable oils by they own are good source of the fat-soluble vitamin. On the other hand, more frequently used vegetable oils for frying decreased in vitamin E content themselves. One of the factor that most influences the vitamin E content in vegetable oils is microwave heating, although the stage of degradation varied with the type of oil used. In addition, vitamin E was unstable in the presence of reducing agents: oxygen, light, and peroxides (occurring as a result of unsaturated fat auto-oxidation). In the referred article, retention of vitamin E was in the range of 44–95% during culinary treatment of various types of meat, and 60–93% in the case of legumes.

Rios & Penteado (2003) studied the effects of ^{60}Co ionizing radiation in doses of 0, 75, 100, 150, 200 and 250Gy on garlic vitamin E content. The α-tocopherol concentration was established by high performance liquid chromatography (HPLC), after direct hexane extraction from the garlic samples and fluorescence detector. It was statistically shown that irradiation dose up to 150 Gy did not affect the garlic α-tocopherol content.

Studies performed at our laboratory (Taipina et al., 2008) aimed at establishing the stability of vitamin E content of industrialized sunflower whole grain cookies after gamma irradiation with doses of 1 and 3 kGy. A high stability of the vitamin content was found but there were significant differences on the sensory characteristics related to appearance, texture, odor and flavor. In another set of experiments the same authors (Taipina et al., 2009) showed that there was no loss of vitamin E content of irradiated pecan nuts up to the dose of 3kGy, although the dose necessary to maintain sensory quality was mere 1kGy. Warner et al. (2008) found that oils with the best oxidative stabilities are those with high γ-tocopherols contents. As already published (Thomas & Gebhardt, 2006) the predominant tocopherol in pecan oil is γ-tocopherol ranging up to 24mg/100g according to that study based on USDA (2006) data. Then, the exact tocopherol composition could be an important factor in the radiation stability found in different studies.

Kim et al. (2009) studied the effect of irradiation on the antioxidant properties of cumin seeds and found that irradiation did not significantly increase and/or maintain all antioxidant parameters.

As it is well established that ionizing radiation stimulate the generation of oxygen radicals which destabilize organic molecules resulting in a decrease of the system's antioxidant potential, Brandstetter et al., (2009) specifically studied the impact of γ-irradiation on the antioxidative properties of some herbs. They found that the effect of irradiation at a dose of 10kGy has little, if any, impact on the antioxidative capacity of the tested herbs.

Skouroliakou et al. (2008) followed the physicochemical stability of prepared admixtures for parenteral nutrition supplied as all-in-one for neonates. Stability assays consisted of the assessment of the all-in-one admixture's (1) macroscopic aspect, (2) drop size measurement, (3) pH measurement, (4) peroxide value, and (5) (x-tocopherol concentration. For the measurements, the admixtures were stored at 2 different temperatures, 4°C (storage) and 25°C (compounding), and then analyzed at a starting time, 24 h, 48 h, and 7 days after compounding. The admixtures for neonates were shown to be physically stable under analysis conditions. The maximum loss of α-tocopherol was approximately 24%. In all-in-one admixtures, lipid peroxide occurred within 24 h after the addition of the lipid emulsion. The addition of fat emulsion and fat-soluble vitamins did not alter the physical stability of parenteral admixtures for neonates. Moreover, the admixtures examined were relatively chemically stable for 24 h, as far as vitamin E was concerned. Lipid peroxidation was the limiting factor for application in terms of stability of an all-in-one neonatal parenteral regimen.

Abd El-Khalek & Taha (2008) conducted a study on 43 male healthy individuals to evaluate the effect of prolonged work in the radiation field on some oxidant-antioxidant status and lipid profile as well. Their results showed that, irrespective of age, occupational radiation exposure caused sustained significant decrease in vitamin E levels while prolonged exposure caused no change in superoxide dismutases and significant increase in malonaldehyde, triglycerides and LDL.

Although considerable work has been conducted on the effects of ionizing radiation on vitamin E in foods and in model systems, available information about radiation products is still insufficient.

Chapter 6

CONCLUSION

Present article was intended to be an overview on the role of vitamin E, a multifunctional substance, in radiation effects expression based on its importance as antioxidant agent. Their antioxidant function is attributed mainly to inhibition of membrane lipid peroxidation and scavenging of reactive oxygen species. Depletion of intracellular antioxidants in acute oxidative stress or in various diseases increases intracellular ROS accumulation. According to Ames theory, the unintended consequences of mechanisms developed during evolution to protect against episodic vitamin/mineral shortages would be responsible for several chronic pathologies including cancer, neurodegenerative or cardiovascular pathologies. Thus, to prevent against cellular damages associated with oxidative stress it would be important to balance the ratio of antioxidants to oxidants by supplementation or by cell induction of antioxidants.

Vitamin E showed to be able to exert protection against radiation damages. From the extended literature on the subject arises the importance to know the exact composition in tocopherols and tocotrienols and whether natural or synthetics vitamers were used in order to draw proper conclusions and to enable the comparison of results.

A great deal of information supports the concept that the stability of vitamin E content as a result of radiation treatment depends mainly on system composition and conditions where irradiation is taking place.

REFERENCES

Abd El-Khalek LG, Taha ME, 2008. The Effect of Prolonged Occupational Radiation Exposure on Lipid Profile and Oxidant-Antioxidant Status. *Arab. J. Nucl. Sc. Applic.* 41(3), 259-268.

Agarwal R, Goel SK, Chandra R, Behari JR, 2010. Role of vitamin E in preventing acute mercury toxicity in rat. *Environ. Toxicol. Pharmacol.* 29(1), 70-78.

Ames BN, 2006. Low micronutrient intake may accelerate the degenerative diseases of aging through allocation of scarce micronutrients by triage. *Proc. Natl. Acad. Sci. U.S.A.* 103, 17589-17594.

Amin AS, 2001. Colorimetric determination of tocopheryl acetate (vitamin E) in pure form and in multi-vitamin capsules. *Eur. J. Pharm. Biopharm.* 51(3), 267-272.

Anon, 1991. Determination of vitamin E in animal feedingsstuffs by high-performance liquid chromatography. *Analyst.* 116 (4), 421-430.

Atay O, Goldal O, Eren V et al., 2009. Effects of Dietary Vitamin E Supplementation on Fattening Performance, Carcass Characteristics and Meat Quality Traits of Karya Male Lambs. *Arch. Animal Breeding.* 52(6), 618-626.

Avula CPR, Fernandes G, 2000. Effects of dietary vitamin E on apoptosis of murine splenocytes. *Nutrit. Res.* 20(2), 225-236.

Biesalski HK, 2009. Vitamin E Requirements in parenteral nutrition. *Gastroenterology.* 137(5), S92-S104.

Borek C, Ong A, Mason H et al, 1986. Selenium and vitamin E inhibit radiogenic and chemically induced transformation in vitro via different mechanisms. *Proc. Natl. Acad. Sci. U. S. A.* 83(5), 490–1494.

Bosch VA, 2005. *La irradiacion de los alimentos.* [Foods irradiation]. Seguridad Nuclear (Madrid) 35: 2-12 .

Bowry VW, Stocker R. 1993. Tocopherol-mediated peroxidation. The prooxidant effect of vitamin E on the radical-initiated oxidation on human low-density lipoprotein. *J. Am. Chem. Soc.* 115, 6029–44.

Brandstetter S, Berthold C, Isnardy B et al., 2009. Impact of gamma-irradiation on the antioxidative properties of sage, thyme, and oregano. *Food Chem. Toxicol.* 47(9), 2230-2235.

Britz SJ, Kremer DF, Kenworthy WJ, 2008. Tocopherols in soybean seeds: Genetic variation and environmental effects in field-grown crops. *J. Am. Oil Chem. Soc.* 85(10), 931-936.

Burton GW, Cheeseman KH, Doba T, Ingold KU, Slater TF, 1983. Vitamin E as an antioxidant in vitro and in vivo. *Ciba Found Symp.* 101, 4–18.

Capuron L, Moranis A, Combe N et al., 2009. Vitamin E status and Quality of Life in the Elderly: Influence of Inflammatory Processes. *Brit. J. Nutrit.* 102(10), 1390-1394.

Cerretani L, Lema-Garcia MJ, Herrero-Martinez JM et al, 2010. Determination of Tocopherols and Tocotrienols in Vegetable Oils byNanoliquid Chromatography with Ultraviolet-Visible Detection Using a Silica Monolithic Column. *J. Agric Food Chem.* 58(2), 757-761.

Chandan KS, Khanna S and Roy S, 2006. Tocotrienols: Vitamin E beyond tocopherols. *Life Sc.* 78(18), 2088-2098.

Chapman JR, Jackson SP, 2008. Phospho-dependent interactions between NBS1 and MDC1 mediate chromatin retention of the MRN complex at sites of DNA damage. *EMBO reports.* 9 (8), 795–801.

Chen X, Touyz RM, Park JB, Schiffrin EL, 2001. Antioxidant effects of vitamins C and E are associated with altered activation of vascular NADPH oxidase and superoxide dismutase in stroke-prone SHR. *Hypertension.* 38(3), 606-611, Part 2.

Choe E, Min DB, 2005. Chemistry and Reactions of Reactive Oxygen Species in Foods. *J. Food Sc.* 70(9), R142-R159.

Chunhieng T, Hafidi A, Pioch D et al, 2008. Detailed study of Brazil nut (*Bertholletia excelsa*) oil microcompounds: Phospholipids, tocopherols and sterols. *J. Brazilian Chem. Soc.* 19(7), 1374-1380.

Clavero MRS, Monk JD, Beuchat LR et al, 1994. Inactivation of Escherichia coli 0157:H7, Salmonellae and Campilobacter jejuni in raw ground beef by gamma irradiation. *Appl. Environ. Microbiol.* 60, 2069-2075.

Crawford LM, Ruff EH, 1996. A review of the safety of cold pasteurization through irradiation. *Food Control.* 7(2), 87-97.

Derr DD, 1993. International regulatory status and harmonization of food irradiation. *J. Food Protect.* 56(10), 882-886.

Diehl JF, 1979. *Einfluss verschiedener bestrahlungsbedigungen und der lagerungen und der lagerung auf strahleninduzierte vitamin E verluste in Lebensmitteln* [Infulence of irradiation conditions and of storage on radiation-induced vitamin E losses in foods]. *Chem. Mikrobiol. Technol. Lebensm.* 6, 65-70.

Diehl JF, 1995. Nutritional adequacy of irradiated foods. In: Safety of Irradiated Foods, 2nd Ed.., New York, NY.Marcel Dekker Inc.

Emmerie A & Engel C, 1938. Colorimetric Determination of dl-α-Tocopherol (Vitamin E). *Nature.* 142(3602), 873-873.

Empey LR, Papp JD, Jewell LD, Fedorak RN, 1992. Mucosal protective effects of vitamin E and misoprostol during acute radiation-induced enteritis in rats. *Dig. Dis. Sci. ;*37(2), 205–214.

Evans, Z.P.; Mandavilli, B.S.; Ellet, J.D. et al., 2009. Vitamin E succinate Enhances Steatotic Liver Energy Status and Prevents Oxidative Damage Following Ischemia/Reperfusion. *Transplantation Proc.* 41(10), 4094-4098.

Fardet A, Rock E, Remesy C., 2008. Is the in vitro antioxidant potential of whole-grain cereals and cereal products well reflected in vivo? *J. Cereal Sc.* 48(2), 258-276.

Farkas J, 1989. Microbiological safety of irradiated foods – Review. *Intl. J. Food Microbiol.* 9, 1-15.

Felemovicius I, Bonsack ME, Baptista ML, Delaney JP, 1995. Intestinal radioprotection by vitamin E (alpha-tocopherol). *Ann. Surg.* 222(4), 504–510.

Franke AA, Murphy SP, Lacey R et al., 2007. Tocopherol and tocotrienol levels of foods consumed in Hawaii. *J. Agric Food Chem.* 55(3), 769-778.

Fryer MJ, 2006. The antioxidant effects of thylakoid Vitamin E (α-tocopherol). *Plant, Cell Environ.* 15(4), 381-892.

Gate L, Paul J, Ba GN, Tew KD, Tapiero H, 1999. Oxidative stress induced in pathologies: the role of antioxidants. *Biomed. Pharmacother.* 53(4), 169-180.

Giardini O, Taccone-Gallucci M, Lubrano R, et al., 1984. Effects of alpha-tocopherol administration on red blood cell membrane lipid peroxidation in hemodialysis patients. *Clin. Nephrol.* 21(3), 174–177.

Goffman FD, Bohme T, 2001. Relationship between fatty acid profile and vitamin E content in maize hybrids (*Zea Mays* L.). *J. Agric Food Chem.* 49(10), 4990-4994.

Goldberg M, Stucki M, Falck J, D'Amours D, Rahman D, Pappin D, Bartek J, Jackson SP (2003) MDC1 is required for the intra-S-phase DNA damage checkpoint. *Nature.* 421, 952–956.

Haflah NHM, Jaarin K, Abdullah S, 2009. Palm vitamin E and glucosamine sulphate in the treatment of osteoarthritis of the knee. *Saudi Med. J.* 30(11), 1432-1438.

Hajimahmoodi M, Oveisi MR, Sadeghi N et al., 2008. Gamma tocopherol content of Iranian sesame seeds. *Iranian J. Pharm. Res.* 7(2), 135-139.

Hall EJ, 1994. Radiobiology for the radiologist. 4th ed. Philadelphia. JB Lippincott Co,

Halliwell B, Gutteridge JMC. 2001. Free radicals in biology and medicine. 3rd ed. New York: Oxford Univ. Press.

Heide L; Boegl KW, 1990. Detection methods for irradiated food - luminescence and viscosity measurements. *Intl. J. Radiat. Biol.* 57, 201-219.

Hemila H, Kaprio J, 2008. Vitamin E supplementation may transiently increase tuberculosis risk in males who smoke heavily and have high dietary vitamin C intake. *British J. Nutrit.* 100(4), 896-902.

Hensley K, Benaksas EJ, Bolli R et al., 2004. New perspectives on vitamin E: γ-tocopherol and carboxyethylhydroxychroman metabolites in biology and medicine. *Free Rad. Biol. Med.* 36(1), 1-15.

Institute of Medicine. 1999- 2001. Food and Nutrition Board. Dietary Reference Intakes. Washington DC, National Academic Press.

Jensen SK & Lauridsen C, 2007. Alpha-Tocopherol stereoisomers. Vitamin E: Vitamins and Hormones Advances in Research and Applications. Book Series 76, 281-308, Elsevier Inc., NY.

Kammerer C, Czermak I, Getoff N, 2001. Radiation protecting properties of Vitamin E-acetate and b-carotene. *Radiat. Phys. Chem.* 60, 71-72.

Khanna, S.;Parinandi, N.L.;Kotha, S.R. et al.,2009. Nanomolar vitamin E α-tocotrienol inhibits glutamate-induced activation of phospholipase A2 and causes neuroprotection. *J. Neurochem.* doi 10.1111/j.1471-4159.2009.06550.x.

Kim H, Kim SG, Choi Y et al, 2008. Changes in tocopherols, tocotrienols, and fatty acid contents in grape seed oils during oxidation. *J. Am. Chem. Soc.* 85, 487-489.

Kim JH, Shin MH, Hwang YJ et al., 2009. Role of gamma irradiation on the natural antioxidants in cumin seeds. *Radiat. Phys. Chem.* 78(2), 153-157.

Knapp FW; Tappel AL, 1961. Comparison of the radiosensitivities of the fat-soluble vitamins by gamma irradiation. *Agric. Food Chem.* 9: 430-433.

Konopacka M, Rzeszowska-Wolny J, 2001. Antioxidant Vitamins C, E and beta-carotene reduce DNA damage before as well as after gamma-ray irradiation of human lymphocytes in vitro. *Mut. Res-Genetic Toxicol. Environ. Mutagenesis.* 491(1-2), 1-7.

Kornsteiner M, Wagner KH, Elmadfa I., 2006. Tocopherol and total phenolics in 10 different nut types. *Food Chem.* 98(2), 381-387.

Kume T, Furuta M, Todoriki S et al., 2009. Status of food Irradiation in the world. *Radiat. Phys. Chem.* 78, 222-226.

Lauridsen C & Jensen SK, 2005. Influence of supplementation of all-rac-{alpha}-tocopheryl acetate preweaning and vitamin C postweaning on {alpha}-tocopherol and immune responses of piglets. *J. Anim. J. Anim. Sci.* 83, 1274-1286.

Lauridsen C, Jensen SK, 2005. Influence of supplementation of all-rac-{alpha}-tocopheryl acetate preweaning and vitamin C postweaning on {alpha}-tocopherol and immune responses of piglets. *J. Anim. J. Anim. Sci.* 83, 1274-1286.

Lawton EJ, Bueche AM, Balwit JS, 1953. Irradiation of Polymers by High-Energy Electrons. *Nature.* 172(4367), 76-77. DOI: 10.1038/172076a0

Leskova E, Kubikova J, Kovacikova E et al., 2006. Vitamin losses: Retention during heat treatment and continual changes expressed by mathematical models. *J. Food Composition Analysis.* 19(4), 252-276.

Lim SJ, Choi MK, Kim MJ, Kim JK, 2009. Alpha-tocopheryl succinate potentiates the paclitaxel-induced apoptosis enforced caspase 8 activation in human H460 lung cancer cells. *Exp. Mol. Med.* 41(10), 737-745.

Lodge JK, 2008. Mass spectrometry approaches for vitamin E research. *Biochem. Soc. Transc.* 36, 1066-1070.

Lucio M, Nunes C, Gaspar D et al, 2009. Antioxidant Activity of Vitamin E and Trolox: Understanding of the Factors that Govern Lipid Peroxidation Studies In Vitro. *Food Biophys.* 4(4), 312-320.

Machlin, L.J., 1991. Handbook of Vitamins, 2nd Ed. New York, M. Dekker.

Mallegol J, Carlsson DJ, Deschenes L, 2001. Post-gamma-irradiation reactions in vitamin E stabilised and unstabilised HDPE. *Nucl. Instr. Meth. Phys. Res. Sect. B-Beam Interact. Mat. Atoms.* 185, 283-293.

Manzi FR. Boscolo FN. Almeida SM. Tuji FM, 2003. Morphological study of the radioprotective effect of vitamin E (dl-alpha-tocopheril) in tissue reparation in rats. *Radiol. Bras.* 36(6), 367-371.

Mastro NLd, 1999a. Development of food irradiation in Brazil. *Progress in Nucl. Energy.* 35(3-4), 229-248.

Mastro NLd, 1999b. Trends in food irradiation in Brazil. In: Book of Synopses. Proceedings of the FAO/AEA/WHO International Conference on Ensuring the Safety and Quality of Food through Radiation Processing, Antalya, Turkey, October 19-22, p. 49-50.

Mastro NLd, Mattiolo SR, 2010. Electron beam irradiation effects on some packaged dried food items. Proceedings of International Topical Meeting on Nuclear ResearchApplications and Utilization of Accelerators, 4-8 May 2009, Vienna, IAEA, Vienna, STI/PUB/1433.

Mastro NLd, Villavicencio ALCH, 1990. *Defesas contra a radiação ionizante em camundongos: ação do alfa-tocoferol e oleo de amendoim.* [Defenses against ionizing radiation in mice: action of alpha-tocopherol and oil peanut]. Short communication. Available from the Nuclear Information Center of Comissao Nacional de Energia Nuclear, RJ (BR). 5th Annual Meeting of the Experimental Biology Societies Federation, Caxambu, MG (Brazil). 24-28 Aug 1990.

McClement DJ, Decker EA, Park Y, Weiss J., 2008. Designing food structure to control stability, digestion, release and absorption of lipophilic food components. *Food Biophys.* 3(2), 219-228.

Meagher EA, Barry OP, Lawson JA et al, 2001.Effects of Vitamin E on Lipid Peroxidation in Healthy Persons. *JAMA.* 285, 1178-1182.

Minkova M, Drenska D, Pantev T, Ovcharov R, 1990. Antiradiation properties of alpha tocopherol, anthocyans, and pyracetam administered combined as a pretreatment course. *Acta Physiol. Pharmacol. Bulg.* 16(4), 31–36.

Minkova M.& Pantev T, 1990. Antiradiation properties of combined pretreatment administration of α tocopherol, anthocyans and pyracetam. *Acta Physiol. Pharmacol. Bulg.* 16(4), 31-37.

Miraliakbari H, Shahidi F, 2007. Lipid class compositions, tocopherols and sterols of tree nut oils extracted with different solvents. *J. Food Lipids.* 15, 81-96.

Miura Y, 2004. Oxidative stress, radiation-adaptive responses, and aging. *J. Radiat. Res.* 45, 357-372.

Miyagi SJ, Brown IW, Chock JML and Collier AC., 2009. Developmental Changes in Hepatic Antioxidant Capacity Are Age- and Sex-Dependent. *J. Pharm. Sci.* 111(4), 440-445.

Miyazaki K, Colles SM, Graham LM, 2008. Impaired graft healing due to hypercholesterolemia is prevented by dietary supplementation with alpha-tocopherol. *J. Vasc. Surg.* 48(4), 986-993.

Oh SY, Chung J, Kim MK et al, 2010. Antioxidant nutrient intakes and corresponding biomarkers associated with the risk of atopic dermatitis in young children. *Eur. J. Clin. Nutr.* doi:10.1038/ejcn.2009.148.

Oliveira A, Rodriguez-Artalejo F, Lopes C, 2009. The association of fruits, vegetables, antioxidant vitamins and fibre intake with high-sensitivity C-reactive protein: sex and body mass index interactions. *Eur. J. Clin. Nutrit.* 63 (11), 1345-1352.

Olson DG, 1998. Irradiation of Food. *Food Technol.* 52(1), 5662.

Oski FA. Vitamin E - a radical defense. *N. Engl. J. Med.* 1980 Aug 21;303(8), 454–455.

Ouanes Z, Abid S, Ayed I et al, 2003. Induction of micronuclei by Zearaleone in Vero monkey kidney cells and in bone marrow cells of mice: protective effect of vitamin E. *Mutation Res. Genetic Toxicol. Environ. Mutagenesis.* 538(1-2), 63-70.

Packer JE, Slater TF, & Willson RL, 1979. Direct observation of a free radical interaction between vitamin E and vitamin C. *Nature.* 278, 737–738.

Pavlik VN, Doody RS, Rountree SD, Darby EJ., 2009. Vitamin E used is associated with improved survival in an Alzheimer's disease cohort. *Dementia Geriat. Cognit. Dis.* 28(6), 536-540.

Rebould E, Thap S, Perrot E et al., 2007. Effect of the main dietary antioxidants (carotenoids, gamma-tocopherol, polyphenols, and vitamin C) on alpha-tocopherol absorption. *Eur. J. Clin. Nutrit.* 61, 1167-1173.

Riley PA, 1994. Free Radicals in Biology: Oxidative Stress and the Effects of Ionizing Radiation. *Int. J. Radiat. Biol.* 65(1), 27-33.

Rios, MDG, Penteado, MVC, 2003. Determinação de □-Tocoferol em alho irradiado utilizando cromatografia líquida de alta eficiência (CLAE). *Quím Nova,* 26(1), 10-12.

Roche M, Tarnus E, Rondeau P, Bourdon E, 2009. Effects of nutritional antioxidants on AAPH- or AGEs-induced oxidativee stress in human SW872 liposarcoma cells. *Cell Biol. Toxicol.* 25(6), 635-644.

Roldi LP, Pereira RVF, Tronchini EA et al, 2009. Vitamin E (α-tocopherol) supplementation in diabetic rats: effects on the proximal colon. *BMC Gastroenterology.* 9(88). doi:10.1186/1471-230X-9-88.

Romero MG; Mendonça AF, 2005. Influence of dietary vitamin E on behavior of *Listeria monocytogenes* and color stability in ground turkey meat following electron beam irradiation. *J. Food Protect.* 68(6), 1159-1164.

Ruperez FJ, Barbas C, Castro M, Martinez S, Herrera E, 1998. Simplified method for vitamin E determination in rat adipose tissue and mammary glands by high-performance liquid chromatography. *J. Chromat. A* 823(1-2), 483-487.

Sarma L, Kesavan PC., 1986. Protective effects of vitamins C and E against gamma-ray-induced chromosomal damage in mouse. *Int. J. Radiat. Biol.* 63(6), 759–764.

Satyamitra M, Devi PU, Murase H, Kagiya VT, 2001. In vivo radioprotection by alpha-TMG: preliminary studies. *Mut. Res-Fund Mol. Mech. Mutagenesis.* 479(1-2), 53-61.

Scherz, H; Senser, F., 2000. Food composition and nutrition tables. Boca Raton, Fl: CRC, 1026.

Schneider, C., 2005. Chemistry and biology of vitamin E. *Mol. Nutrit. Food Res.* 49(1), 7-30.

Sezen O, Ertekin M, Demircan B et al., 2008. Vitamin E and l-carnitine, separately or in combination, in the prevention of radiation-induced brain and retinal damages. *Neurosurgical Rev.* 31(2), 205-213.

Shadyro OI, Sosnovskaya AA, Edimecheva IP, et al., 2005. Effects of various vitamins and coenzymes Q on reactions involving α-hydroxyl-containing radicals. *Free Rad. Res.* 39(7), 713-718.

Shin JH, Jeong SG, Han GS et al., 2008. Reduction of the antigenicity of powdered milk by gamma irradiation. *Korean J. Food Sc. Animal Resources.* 28(3), 306-311.

Shireen KF, Pace, RD, Mahboob M, Khan AT, 2008. Effects of dietary vitamin E, C and soybean oil supplementation on antioxidant enzyme activities in liver and muscles of rats. *Food Chem. Toxicol.* 46(10), 3290-3294.

Singh RK, Verma NC, Kagiya VT, 2001. Effect of a water soluble derivative of alpha-tocopherol on radiation response of *Saccharomyces cerevisiae. Indian J. Biochem. Biophys.* 38(6), 399-405.

Skouroliakou M, Matthaiou C, Chiou A et al., 2008. Physicochemical stability of parenteral nutrition supplied as all-in-one for neonates. *J. Parent Enter. Nutrit.* 32(2), 201-209.

Slater, TF, ed., 1978. In Biochemical Mechanisms of Liver Injury. London, Academic Press, , p. 745–801.

Slatore CG; Littman, AJ, Au DH et al., 2008. Long-Term Use of Supplemental Multivitamins, Vitamin C, Vitamin E, and Folate Does Not Reduce the Risk of Lung Cancer. *Am. J. Resp. Crit. Care Med.* 177, 524-530.

Srinivasan V, Weiss JF, 1992. Radioprotection by vitamin E: injectable vitamin E administered alone or with WR-3689 enhances survival of irradiated mice. *Int. J. Radiat. Oncol. Biol. Phys.* 23(4), 841–845.

Sundl I, Murkovic M, Bandoniene D et al, 2007. Vitamin E content of foods from: Comparison of results obtained from food composition tables and HPLC analysis. *Clin. Nutrit.* 26 (1), 145-153.

Szodoray P, Horvath IF, Papp G et al, 2010. The immunoregulatory role of vitamins A, D and E in patients with primary Sjogren's syndrome. *Rheumatology.* 49(2), 211-217.

Szymanska R, Kruk J, 2008. Tocopherol content and isomers'composition in selected plant species. *Plant Physiol. Biochem.* 46, 29-33.

Taipina MS, Lamardo LCA, Rodas MAB, Mastro NLd, 2008. Vitamin E content and sensory qualities of γ-irradiated sunflower whole grain cookies. *Nukleonika.* 53 (2), S81-S84.

Taipina MS, Lamardo LCA, Rodas MAB, Mastro NLd, 2009. The effects of gamma irradiation on the vitamin content and sensory qualities of pecan nuts (Caya illinoensis). *Radiat. Phys. Chem.* 78, 611-613.

Thayer DW, 1990. Food irradiation: benefits and concerns. *J. Food Qual.* 13, 147-169.

Thomas, R.G., Gebhardt, S.E. 2006. Nuts and seeds as sources of alpha and gamma tocopherols. ICR/WCRF International Research Conference, July 13-14, 2006, Washington, D.C.

Thorne S (ed), 1991. Food Irradiation. Essex, England, Elsevier Sc Publishers Ltd.

Toumpanakis D, Karatza MH, Katsaounou P et al., 2009. Antioxidant Supplementation Alters Cytokine Production from Monocytes. *J. Interferon Cytokine Res.* 29(11), 741-748.

USDA. Nutrient Database for Standard Reference. Release 19 (2006). Nuts pecan. http://www.nal.usda.gov/fnic/foodcomp/search/;iten:pecan;"nuts, pecan".

Uysal T, Amasyali M. Olmez H, Gunhan O, 2009. Stimulation of boné formation in the expanding inter-premaxillary suture by vitamin E in rats. *Korean J. Orthod.* 39(5), 337-347.

Vasilaki AT, Leivaditi D, Talwar D, et al., 2009. Assessment of vitamin E status in patients with systemic inflammatory response syndrome: Plasma, plasma corrected for lipids or red blood cell measurements? *Clin. Chim. Acta.* 409(1-2), 41-45.

Villaverde C, Cortinas L, Barroeta AC et al, 2004. Relationship between dietary unsaturation and vitamin E in poultry. *J. Animal Phys. Animal Nutrit.* 88 (3-4), 143-149.

Wagner KH, Kamal-Eldin A, Elmadfa I, 2004. Gamma-tocopherol- An underestimated vitamin? *Ann. Nutr. Metab.* 48, 169-188.

Warner, K. Miller, J. and Demurin, Y., 2008. Oxidative stability of crude mid-oleic sunflower oils from seeds with γ- and δ-tocopherol levels. *J. Am. Chem. Soc.* 85, 529-533.

Webster, AM. Webter's Third New International Dictionary. USA, p. 2403, 1976.

Weiss JF, 1997. Pharmacologic approaches to protection against radiation-induced lethality and other damage. *Environ. Health Perspect.* 105(Suppl 6), 1473–1478.

Weiss JF, Landauer MR, 2000. Radioprotection by antioxidants. Reactive oxygen species: from radiation to molecular biology, 899, 44-60.

WHO, 1994. Safety and nutritional adequacy of irradiated food, Geneva, World Health Organization.

Woods RJ; Pikaev AK, 1994. Interaction of radiation with matter. In: *Applied Radiation Chemistry: Radiation processing.* John Willey & Sons, New York.

Yanardag R, Bolkent S, Kizir A, 2001. Protective effects of DL-alpha-tocopherol acetate and sodium selenate on the liver of rats exposed to gamma radiation. *Biol. Trace Elem. Res.* 83(3), 263-273.

Yang H., Mahan DC, Hill DA, Shipp, TE, Radke TR., Cecava MJ, 2009. Effect of vitamin E source, natural versus synthetic, and quantity on serum and tissue alpha-tocopherol concentrations in finishing swine. *J. Animal. Sc.* 87(12), 4057-4063.

Yang J, 2009. Brazil nuts and Associated Health Benefits: A review. LWT-*Food Sci. Technol.* 42(10), 1573-1580.

Yong LC, Petersen MR, Sigurdson AJ et al., 2009. High dietary antioxidant intakes are associated with decreased chromosome translocation frequency in airline pilots. *Am. J. Clin. Nutrit.* doi:10.3945/ajcn.2009.28207.

Yoshida Y, Saito Y, Jones LS, Shigeri Y., 2007. Chemical reactivities and physical effects in comparison between tocopherols and tocotrienols: physiological significance and prospects as antioxidants. *J. Biosc. Bioeng.* 104(5), 439-445.

Yoshimura M, Kashiba M, Oka J et al, 2002. Vitamin E prevents increase in oxidative damage to lipids and DNA in liver of ODS rats given total body X-ray irradiation. *Free Rad. Res.* 36(1), 107-112.

Yu WP, Jia, l, Park SK, Li J et al., 2009. Anticancer Actions of Natural and Synthetic Vitamin E Forms: RRR-Alpha-Tocopherol Blocks the Anticancer Actions of Gamma-tocopherol. *Molec. Nutrit. Food Res.* 53(12), 1573-1581.

INDEX

S

T